Decorative and Ornamental
BRICKWORK

162 Photographic Illustrations

JAMES STOKOE

Dover Publications, Inc., New York

For Caroline

Acknowledgments: Washington University School of Architecture; Missouri Historical Society; Margaretta Darnall; Elmer Kempf; Allan Stock; Richard Bennett Lord.

Published in Canada by General Publishing Company, Ltd., 30 Lesmill Road, Don Mills, Toronto, Ontario.
Published in the United Kingdom by Constable and Company, Ltd., 10 Orange Street, London WC2H 7EG.

Decorative and Ornamental Brickwork: 162 Photographic Illustrations is a new work, first published by Dover Publications, Inc., in 1982.

International Standard Book Number: 0-486-24130-0
Library of Congress Catalog Card Number: 81-70897

Manufactured in the United States of America
Dover Publications, Inc.
180 Varick Street
New York, N.Y. 10014

INTRODUCTION

This is a collection of photographs about the use of brick for the decoration of architecture. It is also an attempt at documenting an aspect of our building heritage that until recently had been largely forgotten.

The first decorative brickwork in America was related to an English tradition. Georgian brickwork employed carved, rubbed and molded shapes to create Palladian classical details *(fig. 1)*. The origins of this brickwork can be traced back to ancient Rome, where, as a detail from a columbarium or funerary structure on the Via Appia Antica illustrates *(fig. 2)*, specially shaped or carved bricks were also made to form classical elements. Other examples of this type of brickwork can be seen in the excavated Roman port of Ostia Antica.

It was not till the end of the Civil War that an American decorative brickwork can be said to have emerged. Interest had grown among such architects as Richard Norman Shaw in England and H. H. Richardson in this country in the possibilities of decorative brick and stone masonry. Brick companies were developing new processes and machinery for brickmaking that would change the product dramatically. The most important process to emerge pressed bricks from a naturally dried surface clay *(fig. 3)*. With this method it was possible to press face brick with the precision and uniformity of cut stone and to form intricate designs.

Fig. 1.

Fig. 3.

Fig. 2.

Taking the new process in a logical direction, the brickmakers revived the practice of making specially shaped bricks and ornamental units based on the brick module. With iron and brass molds and steam-driven presses, companies rapidly built up inventories of special bricks. Their designs reflected contemporary interest in plant forms, which could be naturalistic representations or examples of stylized flower and leaf forms

Fig. 4.

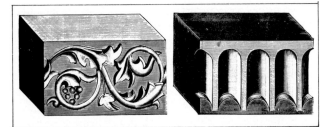

Fig. 6.

Fig. 7.

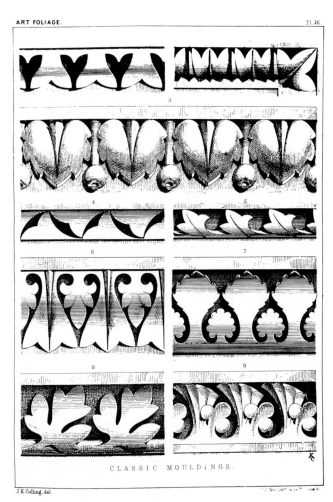

CLASSIC MOULDINGS.

J.K. Colling. del

Fig. 5.

from Japanese or Egyptian architecture *(fig. 4)*. Other designs were derived from naturalizing classical forms of ornament *(fig. 5)*.

Catalogs grew to include almost every known historical ornament. There were Georgian moldings, Tudor-like tracery designs, classical forms and rounded and ropework shapes from sixteenth-century German brickwork. The brickmakers' catalogs were no less than lexicons of ornamental forms *(figs. 6 and 7)*.

These bricks were not ornaments to be applied to a surface as much as they were part of a system of ordered and interchangeable parts *(fig. 8)*. Many of the designs, such as projections and hinge bricks, made the mason's job easier and faster. The brick had become a precise machine-made product and has to be included as part of the emergence of the nineteenth-century "machine aesthetic" that has been attributed to balloon framing and metal-skeleton structures.

Because of the accessibility of these products and the easy manipulation of the units, brick became a kind of consumer vernacular allowing the layman to express, with the collaboration of the mason, a great deal of creative spirit. This architecture was as exuberant and inventive as any of the many wood vernacular styles *(fig. 9)*.

Around the turn of the century, however, the architectural profession began to look on the use of manufactured ornamental bricks with discomfort. Coinciding with increased European travel by young architects and the formation of architectural associations in many cities, polemics on the proper use and manufacture of bricks began to appear. The artificial character of pressed brick came under attack. Examples of Italian medieval architecture were argued as models to be followed *(fig. 10)*.

In the next 15 years the brickmakers responded to the calls for a more natural product. They developed new techniques for producing wide ranges of color and new machinery that scratched, scored and patterned bricks so the manufacturers could claim their bricks were "inspired by the irregular and intimate quality of things made entirely by hand."

There were no longer integral decorative units in a system. Decorative forms came from external sources,

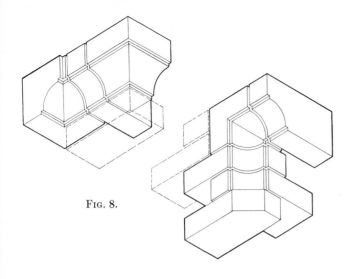

Fig. 8.

Fig. 10.

Fig. 9.

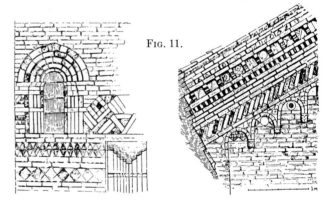

Fig. 11.

first from medieval examples, but often from an analogy between brickwork and woven material. Manufacturers marketed their products with such names as: Afghan Vertical Ruffs, Full Range Velour, Rug Shades, Persians and Chenilles. Pictorial examples in periodicals often stressed textural patterned motifs *(fig. 11)*.

The brick decoration of the first half of the twentieth century was much more the work of professionals. Yet vernacular work did occur because the brick companies' advertising popularized certain forms. Professional or nonprofessional, there was invention and play with texture and patterns that gives the embellished structures of these decades not the radical expressive spirit of the nineteenth century, but a worldly richness.

By the end of the Second World War there was a general acceptance of the tenets of the modern movement among architects. This meant the end of applied decoration. Brick was considered only as a uniform texture. But equally as numbing as the modern style was the scale of residential speculation. The examples in this book for the most part represent a tradition of owner-

built homes and small-scale speculation. In the late forties and fifties, with the trend toward larger projects, the production of housing had to simplify formal decisions out of necessity. Only a handful of people made the choices, whereas 50 years before, hundreds of individuals were involved. The tendency away from decoration was agreeable to the developer. Large parts of our environment have been shaped by this expedient.

We are fortunate that our cities, despite neglect, still retain examples of decorative traditions. Decorative brickwork still lines streets in such cities as Chicago, Washington, Philadelphia, New York and Boston. It is in St. Louis, however, that there seems to be the most complete statement. Most examples have been taken from this city, a city that troweled its way into the twentieth century with a hopeful art and vigor.

CONTENTS

THE CORNICE

The cornice—the crowning projection of a wall—was an extremely popular embellishment of nineteenth-century structures. Aesthetically, the cornice celebrated the transition between man's endeavors and the sky. More pragmatically, the cornice was used to give a structure a grandeur commensurate with the expectations of the era in which it was built. By projecting from the top of the wall, perspective could be distorted and verticality enhanced.

In a large building the successive projecting or corbeling of bricks might take a couple of stories to form a cornice. A strategy employed in many late nineteenth-century structures in Washington, D.C., used beveled corners to create corbeling that otherwise would not have been necessary *(page 5, top left and right)*.

In all corbeling there is a limited distance which each successive brick can extend beyond the one below if structural stability is not to be forfeited. What was needed (and was in fact manufactured by brick companies) was longer or projecting shapes that could maintain stability while creating cornices in fewer courses. Yet, with catalogs of all sorts of shapes and ornaments to choose from, the issue quickly shifted from the utility of projections to the possibilities of combining elements to create cornices expressing complex shadings of grandeur.

Looking at the results of this catalog vernacular, an academically trained architect might bemoan the garbling of the classical language in which the architectural orders are defined in part by a specific stacking of the correct projecting forms (many of which could be found in the catalogs). Yet the free invention on the theme of the classical cornice is the key to the richness of this nineteenth-century tradition.

The brick manufacturers also made larger-scale ornamental units, the largest being a brick 10″ x 10″ x 2⅜″. Because of the bricks' size and square shape, they could be pressed into intricate centralized designs. These units have ancestors in Greek friezes such as those that survive in fragments at Paestum. A more contemporary source was the sunflower forms adapted from oriental designs by English architects such as W. E. Nesfield.

The centralized brick designs were executed in naturalistic florals or abstractions expressing the mechanized origins of the product. Expanding the potential of the 10″ x 10″ brick, manufacturers produced designs that formed larger or continuous patterns when combined. They also made 5″ x 10″ x 2⅜″ bricks. Some were designed to be used in pairs to make 10″ x 10″ figures, but many were used for larger-scale decorative bands. They also made 5″ x 5″ x 2⅜″ bricks which were used analogously to the larger units, but which because of their smaller size allowed more flexibility of location and more room for repetitive patterns.

In contrast to the cornice embellishments of the 1880s and 1890s, after the turn of the century bricks and buildings ceased to project at the cornice, corbeling being done in a much more painterly manner. Distinct shadows were no longer desired. In fact, there is a rich progression to a demarcation of the top of the wall with pattern instead, certainly a no less pleasing and also largely ignored possibility of decorative brick.

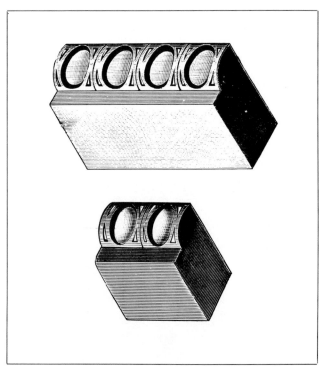

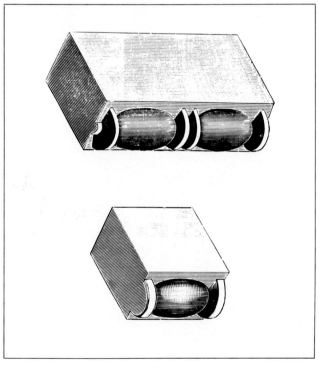

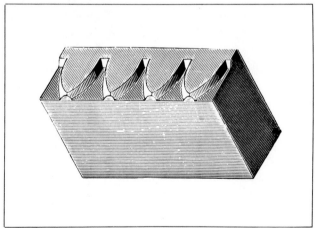

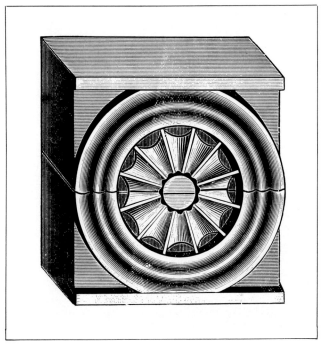

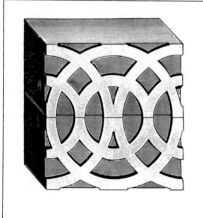

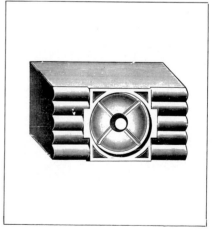

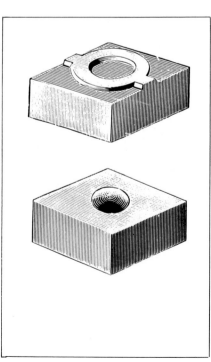

GABLES AND DORMERS

The gable was sometimes ornamented as a sloping cornice with corbeling, consistent with the treatment of the classical pediment. More frequently, however, the brickwork reflects the structural redundancy of the portion of the wall between the sloped sides of the gable. (In early Etruscan temples those spaces were left open; the Romans often placed large window openings there.)

In the brick gable walls of the late nineteenth century we find openings or groups of multiple openings. Gables often enclosed attic spaces too small to require windows. In these examples the use of the 5″ x 5″ bricks in stacking repetitive patterns seems like a mimicry of the nineteenth-century revival of tile hanging in England. The effect of the surface created by these simple shapes is, however, like nothing in the English vernacular tradition, the American surfaces being straightforward expressions of machine-aided ornamental architecture. But different designs and sizes of bricks made by the same machines could synthesize a surface with seemingly historical richness *(page 36)*.

Sometimes the gables themselves are shams, having no rooms behind them. In these instances ornamental bricks were used to create panels where in a real gable a window might have appeared. Dormers, however, had real windows providing light to rooms that extended into the roof. The face of the dormer was pulled forward and literally grew out of the cornice. So the investment in a brick dormer could give a small house extra height and importance. Column forms made from various special-shaped bricks abound as a way to give visual stability to these precarious upward extensions.

The approach to these elements, when the shapes and ornaments had disappeared and the colors and textures were the mason's materials, was not fundamentally different. The nonarchitectural associations of weaving patterns lent themselves well to the small structural demands of the gable ends. A fascination with the diagonal and zigzag patterns began to appear in the form of slanted coursing. A special pattern used in gable ends and walls was the diaper, which was popularized by architects in this country who favored its medieval origins. It is a diagonal pattern formed by staggering header bricks, or bricks of color or texture contrasting with those used in the rest of the wall.

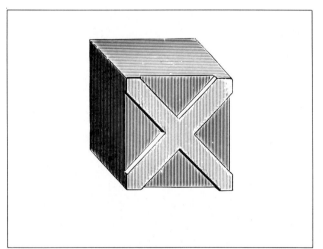

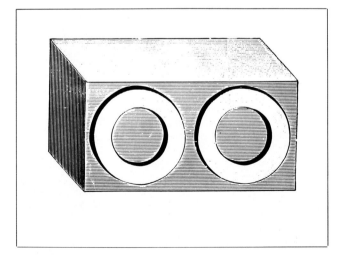

DOORS AND WINDOWS

Next to the cornice, the portions of the late nineteenth-century facade that received the most attention were the windows and doorways. To have a grand house required a doorway that expressed the expectations of the people who lived within.

Variety was the rule, but the basic opening was the arch. (Today the cost of skilled labor makes the flat steel lintel less expensive.) The arch is an economical structural use of brick. The technology of brickmaking at that time made it easier for the mason to form the arch in a precise and structurally sound manner. The dense dry-press bricks were ground into wedge shapes for arches of specific radii. In a large arch the form could be made by creating wedge-shaped mortar joints. Shaping the bricks made arches of smaller radius possible.

Flat-topped openings often employed systems of bricks that created frames around them. The arched opening, however, because of the changing structural roles of the bricks from the top to the sides, had opportunities for decoration or articulation not present in the flat-topped opening. First the arch bricks were separated from the horizontal coursing of the wall by a banding of a repeated decorative shape or the arch bricks themselves were ground from contrasting ornaments. At the base of the arch there was often an im-

post or simplified capital which articulated the change from arch bricks to the bricks that formed the sides of the openings. These in turn might be marked as different from the wall. A popular device was the use of special shapes, as in the dormers, to form column-like articulations.

Other historically based areas for ornament were the spandrels, the roughly triangular panels at the side of the arch, and the tympanum, the area formed when the top of the arch is filled in *(pages 64 and 65)*.

An unusual location of ornament can be seen in the top illustration on page 66, where four 10″ x 10″ bricks form one large centralized pattern. It is also noteworthy because the mason or owner has stumbled onto what was the principal ornamentation of the window in the early twentienth century.

The space below one window and above another, also called the spandrel, was an ideal location for the free patterning of brick in diagonal and basket-weave forms.

When special importance was desired for an opening the stepping form was sometimes employed. It was "drawn" with contrasting bricks or the opening itself was stepped. In this period there was also much more experimentation with combining other materials, such as terra-cotta, stucco or colored glass panels, with brickwork.

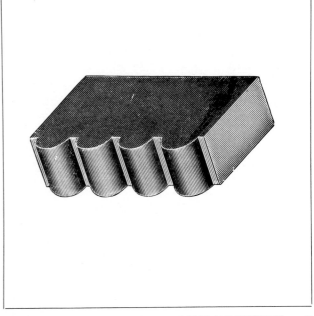

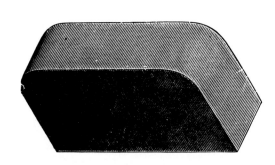

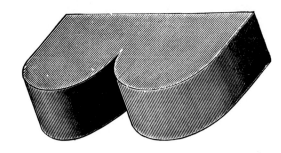

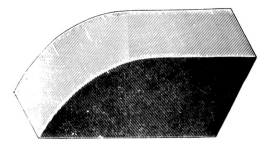

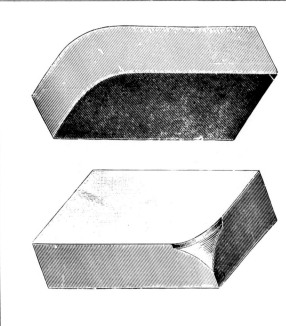

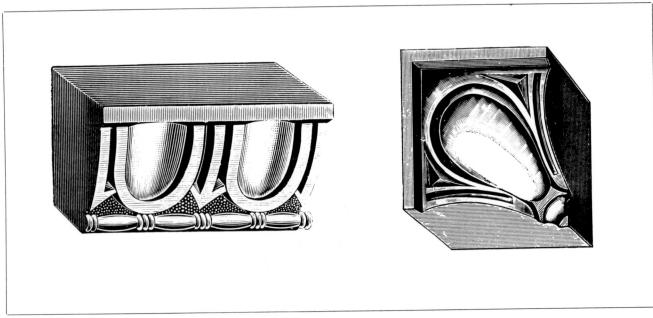

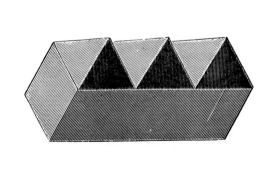

THE WALL

In the early twentieth-century revolt against the precision of the face brick, the wall itself took on many different characters. The bricks, the mortar joints and the coursing were all subject to variation. The resulting patterns or textures were then spread out over large areas of wall surface. Rather than incorporate a few ornamental shapes to decorate a cornice in a house, a specially patterned brick might be used for everything. These bricks had patterns made by contrasting rough and smooth areas on the same brick. There was a brick that had a scored line dividing it in halves; one side being smooth, the other rough. The effect was an illusion of tile or of smaller, more intricate units *(page 77, top)*.

When two or more different styles of brick were used in a wall, they could be mixed at random, but more often they were organized in a pattern such as the diaper or in stylized versions of other two-dimensional motifs. Banding or striping of the wall with a contrasting brick was a popular form of embellishment which could emphasize the horizontal form of a modern building. It was also timely that these patterns often paid little attention to the articulation of the traditional parts of a

facade. Architects were moving toward an architecture without ornament. Yet, along the way, there were extremely rich inventions.

The late nineteenth-century brick wall was a hard smooth surface of uniform color. The principal ornamentations of the wall were string courses which helped give scale and compose the facade by relating windows to common base lines. Regular-size bricks with projecting designs were often used, but the 5″ x 10″ brick designs were more effective because of their large scale. 5″ x 5″ bricks were also used.

Occurring much less often than the string course was the panel, a very special use of decorative brick shapes. One of the finest examples, H. H. Richardson's Trinity Church Rectory in Boston, is actually an example of carved brick. Yet the panels made from the catalog shapes are, in a way, more interesting. They are small celebrations of the products of that machine age. They are not so much exciting as isolated events but as being part of a building fabric that is so susceptible to creative variations.

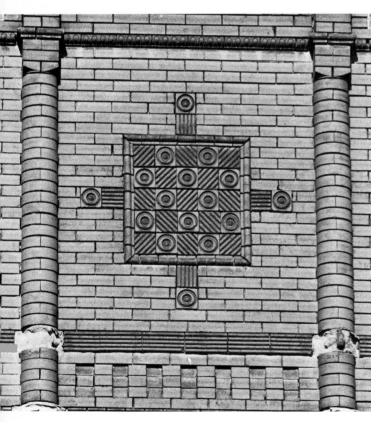